INTERNATIONAL CENTRE FOR MECHANICAL SCIENCES

COURSES AND LECTURES - No. 94

HEINZ PARKUS
TECHNICAL UNIVERSITY OF VIENNA

OPTIMAL FILTERING

COURSE HELD AT THE DEPARTMENT
FOR GENERAL MECHANICS
SEPTEMBER 1971

UDINE 1971

SPRINGER-VERLAG WIEN GMBH

This work is subject to copyright.

All rights are reserved,

whether the whole or part of the material is concerned

specifically those of translation, reprinting, re-use of illustrations,

broadcasting, reproduction by photocopying machine

or similar means, and storage in data banks.

Copyright 1972 by Springer-Verlag Wien

Originally published by Springer—Verlag Wien—New York in 1972

ISBN 978-3-211-81130-6   ISBN 978-3-7091-2886-2 (eBook)
DOI 10.1007/978-3-7091-2886-2

PREFACE

This booklet is intended to serve as a text for my lectures given at the International Centre for Mechanical Sciences in Udine during the first half of October, 1971.

The problem of controlling a system if measurements of its state and, perhaps, the state itself are subjected to random disturbances has created an urgent need for effective filters to eliminate these disturbances. Since the pioneering work of Norbert Wiener an enormous amount of research has been done in this field. I hope, however, that I have succeeded in giving, in this short text, an introduction which, although brief, nevertheless touches upon the essential aspects of the subject. The book by Bryson and Ho, written explicitly for engineers, provides an excellent vehicle for further study. For numerous practical applications in aeronautics and astronautics the book by Greensite may be consulted. For concepts and formulas from the theory of stochastic processes, necessary for an understanding of the text, the reader is referred to my book "Random Processes in Mechanical Sciences".

It is with great pleasure that I record here my sincere thanks to Prof. Luigi Sobrero, Secretary

General of CISM, and to Prof. *Waclaw Olszak*, Rector of CISM, for making it possible for me to participate in the activities of the Centre.

H. Parkus

Introduction

In order to control a system effectively in an optimal manner a knowledge of the state of the system at every instant of time is required. This information, however, can only be obtained by observing the system, i.e. by taking measurements, continuously or at discrete instants of time, of the state variables (e.g. position, velocity, temperature etc.) or, more frequently, of some functions of these variables. In practice, one finds then that the observations always contain random errors. In addition, the system may itself be subjected to random disturbances. Also, at times, we may have insufficient information, while on other occasion, we may have more than enough measurements, so that some or all state variables are overdetermined. In all these circumstances, we are not able to give the precise values of the state variables but are forced to make an estimate. Of course, we wish to make this estimate as good as possible, in some well-defined sense. This optimization problem will be discussed in the following.

The first to attack this problem was <u>Norbert Wiener</u>. In his famous book "The Extrapolation, Interpolation, and Smoothing of Stationary Time Series", New York 1949, he laid the groundwork for all subsequent research in this field. His solution will be discussed in Chapter I. Later on, in 1960 and

1961, R.E.Kalman and R.S.Bucy tackled the problem from an entirely different point of view, thereby generalizing the results of Wiener. Their work will be discussed in Chapter II. Both the Wiener and the Kalman-Bucy filter are applicable to linear systems only. Some possibilities of using them in connection with nonlinear systems will be discussed in Chapter III. The final Chapter IV presents a very brief introduction into the technique of using optimal filtering for optimal control.

Chapter 1

THE WIENER FILTER

We consider a linear, open loop, deterministic system.

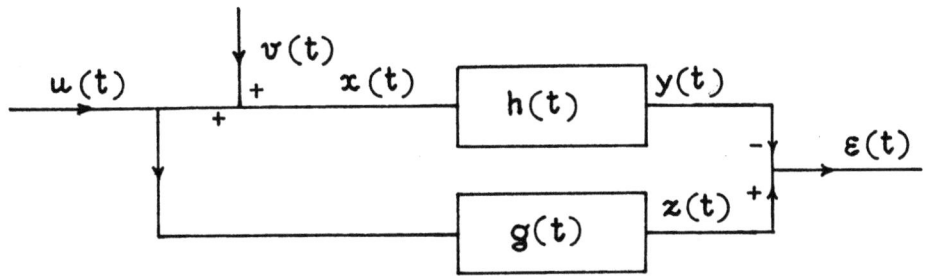

The ideal system $g(t)$ transforms the undisturbed random test signal $u(t)$ into $z(t)$. The actual system or filter $h(t)$, to be optimized, transforms the distorted signal $x(t) = u(t) + v(t)$, where $v(t)$ is random noise, into $y(t)$. We wish to have $y(t)$ "as close as possible" to $z(t)$ by minimizing the mean square error (dispersion D):

$$D = E\{\varepsilon^2(t)\} = E\{[z(t) - y(t)]^2\} = \min.$$

Since we want to eliminate the perturbation noise $v(t)$ from $y(t)$ we speak of <u>filtering</u>.

If $z(t) = u(t+\tau)$, we have the case of <u>extrapolation</u> or <u>prediction</u> where the future values of $u(t)$ are to be predicted from those of the past.

If

$$z(t) = \frac{d^k u(t+\tau)}{dt^k}$$

we have the case of <u>differentiation</u>.

1.1. The Wiener-Hopf Equation

We assume the random processes $u(t)$ and $v(t)$ to be stationary with zero mean and known autocorrelation and cross-correlation. Furthermore, we assume that $x(t)$ has been observed through an unlimited period $(-\infty, t]$ preceding, and up to, time t.

Using the impulse response $h(t)$ of the filter we have for the dispersion, see [1], Eq. (I - 5.20),

$$(1.1) \quad D = E\left\{\left[\int_0^\infty h(\tau)x(t-\tau)d\tau - z(t)\right]^2\right\}$$

or, upon squaring,

$$(1.2) \quad D = \int_0^\infty h(s)\left[\int_0^\infty h(\tau)R_{xx}(s-\tau)d\tau - 2R_{zx}(s)\right]ds + R_{zz}(0)$$

where $h(t)$ has to be determined such as to make D a minimum. To solve this variational problem we replace $h(t)$ by the expression

$$h(t) + \alpha\eta(t)$$

and require

$$\left.\frac{\partial D}{\partial \alpha}\right|_{\alpha=0} = 0$$

Solution of the Wiener-Hopf Equation

whence

$$\int_0^\infty \eta(s) \left[\int_0^\infty h(\tau) R_{xx}(s-\tau) d\tau - R_{zx}(s) \right] ds = 0.$$

From the fundamental lemma of the calculus of variations we get

$$\int_0^\infty h(\tau) R_{xx}(t-\tau) d\tau - R_{zx}(t) = 0 \qquad (1.3)$$

for $t \geq 0$, corresponding to the interval of integration of s.

Eq. (1.3) is known as the <u>Wiener-Hopf integral equation</u> for the optimal impulse response $h(t)$. Due to the fact that it is valid for nonnegative t only, its solution is not straightforward.

If (1.3) is substituted into (1.2) one obtains for the minimum dispersion

$$D_{min} = R_{zz}(0) - \int_0^\infty h(s) R_{zx}(s) ds . \qquad (1.4)$$

1.2. Solution of the Wiener-Hopf Equation

We first transform to spectral densities using [1], Eqs. (I - 5.17), and introduce the frequency response

$$F(\omega) = \int_0^\infty h(\tau) e^{-i\omega\tau} d\tau .$$

Eq. (1.3) then goes over into

$$\int_{-\infty}^{+\infty} [F(\omega) S_{xx}(\omega) - S_{zx}(\omega)] e^{i\omega t} d\omega = 0 \text{ for } t \geq 0. \qquad (1.5)$$

To solve this integral equation for the optimal frequency response $F(\omega)$ we employ the following results from complex function theory:

Let $\Phi^+(\zeta)$ be a function without poles in the upper half complex plane $\eta \geq 0$, where $\zeta = \xi + i\eta$, and let $\Phi^-(\zeta)$ be a function without poles in the lower half-plane $\eta \leq 0$. Then

(1.6)
$$\begin{cases} \int_{-\infty}^{+\infty} \Phi^-(\xi) e^{i\xi t} d\xi = 0 & \text{if} \quad t < 0 \\ \int_{-\infty}^{+\infty} \Phi^+(\xi) e^{i\xi t} d\xi = 0 & \text{if} \quad t \geq 0. \end{cases}$$

The proof may easily be given using the calculus of residues and closing the integration path by semicircles, with $R \to \infty$, in the lower and upper half-planes, respectively. A condition on $\Phi(\zeta)$ is appropriate behavior for $|\zeta| \to \infty$. Since it will be assumed in the following that $\Phi(\zeta)$ can be represented as a Fourier integral,

(1.7)
$$\int_{-\infty}^{+\infty} |\Phi(\xi)|^2 d\xi < \infty$$

will be valid, and the condition will be satisfied.

The second of Eqs. (1.6) is identical with Eq. (1.5). Hence, the function

(1.8)
$$\Phi^+(\zeta) = S_{zx}(\zeta) - F(\zeta) S_{xx}(\zeta)$$

must not have any poles in the upper half-plane.

We assume the optimal system to be stable. There-

Solution of the Wiener-Hopf Equation

fore, its transfer function $Y(s)$,

$$Y(s) = \int_0^\infty h(t) e^{-st} dt$$

cannot have poles with positive real parts of s. But, since $F(\omega) = Y(i\omega)$, it follows that $F(\omega)$ cannot have poles with negative imaginary parts of ω, $F(\omega) = F^-(\omega)$. Eq. (1.8) may now be written as

$$\Phi^+(\zeta) + F^-(\zeta) S_{xx}(\zeta) = S_{zx}(\zeta). \qquad (1.9)$$

Spectral density $S_{xx}(\zeta)$ will, in general, have poles both in the upper and lower half-plane. In many instances, however, it is possible to represent S_{xx} as the product of a function $S^+(\zeta)$, without poles and zeros in the upper, and a function $S^-(\zeta)$, without poles and zeros in the lower half-plane:

$$S_{xx}(\zeta) = S^+(\zeta) S^-(\zeta). \qquad (1.10)$$

This can be done in a particularly simple manner, if S_{xx} is the quotient of two polynomials (*).

Eq. (1.9) may now be written as

$$\frac{S_{zx}(\zeta)}{S^+(\zeta)} = \frac{\Phi^+(\zeta)}{S^+(\zeta)} + F^-(\zeta) S^-(\zeta). \qquad (1.11)$$

We have thus resolved the function $S_{zx}(\zeta)/S^+(\zeta)$ into two parts,

---

(*) For a more general case, see [2], p. 136.

the first representing the value, along the real axis, of a function analytic in the upper half plane, and the second representing the value, along the real axis, of a function analytic in the lower half plane. Now, such a resolution may, however, also be obtained with the aid of the Fourier transform. Indeed, for a function $\Phi(\xi)$ satisfying (1.7),

$$\Phi(\xi) = \frac{1}{2\pi} \int_{-\infty}^{+\infty} e^{-i\xi t} dt \int_{-\infty}^{+\infty} \Phi(\omega) e^{i\omega t} d\omega = \frac{1}{2\pi} \left[\int_{-\infty}^{0-} e^{-i\xi t} dt + \int_{0}^{+\infty} e^{-i\xi t} dt \right] \int_{-\infty}^{+\infty} \Phi(\omega) e^{i\omega t} d\omega$$

whence at once

(1.12)
$$\begin{cases} \Phi^-(\xi) = \dfrac{1}{2\pi} \int_{0}^{+\infty} e^{-i\xi t} dt \int_{-\infty}^{+\infty} \Phi(\omega) e^{i\omega t} d\omega \\[2ex] \Phi^+(\xi) = \dfrac{1}{2\pi} \int_{-\infty}^{0-} e^{-i\xi t} dt \int_{-\infty}^{+\infty} \Phi(\omega) e^{i\omega t} d\omega \ . \end{cases}$$

For a proof it suffices to substitute Eqs. (1.12) into (1.6), and to use the delta function representation

$$\frac{1}{2\pi} \int_{-\infty}^{+\infty} e^{i(t-\tau)\xi} d\xi = \delta(t-\tau)$$

together with the relation

$$\int_{0}^{\infty} e^{i\omega\tau} \delta(t-\tau) d\tau = 0 \qquad \text{for} \qquad t < 0 \ .$$

Solution of the Wiener-Hopf Equation

Upon replacing now the left-hand side of Eq. (1.11) by its Fourier transform, and equating the − parts on both sides one obtains

$$F(\omega) = \frac{1}{2\pi S^-(\omega)} \int_0^\infty e^{-i\omega t} \int_{-\infty}^{+\infty} \frac{S_{zx}(\alpha)}{S^+(\alpha)} e^{i\alpha t} \, d\alpha \, dt \, . \qquad (1.13)$$

This equation determines the frequency response of the optimal filter.

Example

We consider the case of **prediction**, i.e. we put

$$z(t) = u(t + \tau) \, . \qquad (a)$$

For $\tau = 0$ we have **filtering**. The spectral densities are assumed as (cf. [1], p. 48),

$$S_{uu}(\omega) = \frac{2\beta}{\beta^2 + \omega^2}, \quad S_{vv}(\omega) = c, \quad \beta, c > 0 \qquad (b)$$

that is, we have white noise disturbances. In addition, we suppose that signal $u(t)$ and noise $v(t)$ are uncorrelated, $R_{uv} = 0$ and hence $S_{uv} = 0$.

From [1], Eq. (II − 2.8) we have

$$S_{xx}(\omega) = S_{uu}(\omega) + S_{vv}(\omega) + S_{uv}(\omega) + S_{vu}(\omega) = S_{uu}(\omega) + S_{vv}(\omega)$$

or, after substitution from Eq. (b),

$$S_{xx}(\omega) = \frac{2\beta + c(\beta^2 + \omega^2)}{\beta^2 + \omega^2} = \frac{(\omega + i\lambda)(\omega - i\lambda)}{(\omega + i\beta)(\omega - i\beta)}$$

where

(c) $$\lambda = \sqrt{\beta^2 + \frac{2\beta}{c}} > 0.$$

Hence, using Eq. (1.10),

(d) $$S^+(\omega) = \frac{\omega + i\lambda}{\omega + i\beta} \quad , \quad S^-(\omega) = \frac{\omega - i\lambda}{\omega - i\beta}.$$

Furthermore, upon multiplying Eq. (a) by $x(t_1) = u(t_1) + v(t_1)$, and taking expectation, we obtain, with $t - t_1 = \mu$,

$$R_{zx}(\mu) = R_{uu}(\mu + \tau) + R_{uv}(\mu + \tau) = R_{uu}(\mu + \tau).$$

Hence

(e) $$S_{zx}(\omega) = \int_{-\infty}^{+\infty} R_{uu}(\mu + \tau) e^{-i\omega\mu} d\mu = e^{i\omega\tau} S_{uu}(\omega).$$

Eq. (1.13) reads now

$$F(\omega) = \frac{\omega - i\beta}{2\pi(\omega - i\lambda)} \int_0^\infty e^{-i\omega t} \int_{-\infty}^{+\infty} \frac{2\beta}{\alpha^2 + \beta^2} \frac{\alpha + i\beta}{\alpha + i\lambda} e^{i\alpha(t + \tau)} d\alpha dt.$$

Solution of the Wiener–Hopf Equation

The inner integral may be evaluated using residues:

$$\int_{-\infty}^{\infty} \frac{2\beta}{(\alpha - i\beta)(\alpha + i\lambda)} e^{i\alpha(t+\tau)} d\alpha = 2\pi i \, \text{Res}(\alpha = i\beta) =$$

$$= 2\pi i \frac{2\beta}{i(\beta + \lambda)} e^{-\beta(t+\tau)}.$$

Then

$$F(\omega) = \frac{2\beta}{\beta + \lambda} \frac{\omega - i\beta}{\omega - i\lambda} e^{-\beta\tau} \int_0^{\infty} e^{-(\beta + i\omega)t} dt = \frac{2\beta}{(\lambda + \beta)(\lambda + i\omega)} e^{-\beta\tau}. \quad (f)$$

This is the frequency response of the desired filter.

Chapter 2
THE KALMAN FILTER

In the preceding chapter we were concerned with finding a linear system that makes the mean square error between actual and desired output a minimum. The result was an integral equation whose solution yielded the impulse response of the required system (filter). Stationary processes and an unlimited observation period were assumed.

The situation becomes considerably more complicated if only finite observation periods, cf. [2], or nonstationary processes are admitted. A more general and much simpler approach has been suggested by <u>Kalman</u> [3] and <u>Kalman and Bucy</u> [4]. This approach is based not on transfer functions and spectral densities of signal and noise as in Wiener's theory but on the difference equations or differential equations, as the case may be, of the system. In other words, the concept of state and state transition is used. After 1960 and 1961, when Kalman introduced the concepts of controllability and observability, it has become clear that the transfer function approach is not fully equivalent to the state variable approach (Kalman-Gilbert theorem).

<u>Note</u>. Essential in the Kalman theory is the assumption that the noise acting on the system, on the measurement, or on both is purely random. If the actual stationary noise does not have this property it is re-

Best Estimate for Linear Systems

placed by random noise generated by passing purely random noise through a linear filter (shaping filter), cf. p. 25, 35.

2.1. Best Estimate for Linear Systems [6]

Let the state of a system be represented by the n-component state vector x. Let measurements of a p-vector m be made, linearly related to x and contaminated by noise v,

$$m = Hx + v \qquad (2.3)$$

where H is a known $(p \times n)$-matrix, and x and v are uncorrelated.

We wish to estimate x. If no measurements were available, some guess will have to be made which we denote by $\bar{x}$. We now employ the measurement to improve on this estimate.

It seems reasonable to use a weighted least square estimate by demanding that the improved estimate, which we denote by $\hat{x}$, minimizes the nondimensional sum

$$J = \frac{1}{2}[(x - \bar{x})^T M^{-1}(x - \bar{x}) + v^T R^{-1} v] \qquad (2.4)$$

The $(n \times n)$-weighting matrix M, defined by

$$M = E\{(x - \bar{x})(x - \bar{x})^T\} \qquad (2.5)$$

is the error covariance matrix <u>before</u> measurement and is assumed to be known. The noise vector v is supposed to have zero expectation and covariance R

$$E\{v v^T\} = R \qquad (2.6)$$

R is a known $(p \times p)$-matrix.

From

$$J = \frac{1}{2}[(x - \bar{x})^T M^{-1}(x - \bar{x}) + (m^T - x^T H^T) R^{-1}(m - Hx)]$$

and

$$dJ = [M^{-1}(x - \bar{x}) - H^T R^{-1}(m - Hx)] dx^T = 0 \text{ for } x = \hat{x}$$

we obtain

$$M^{-1}(\hat{x} - \bar{x}) - H^T R^{-1}(m - H\hat{x}) = 0$$

or, after rearranging,

$$(M^{-1} + H^T R^{-1} H)\hat{x} = (M^{-1} + H^T R^{-1} H)\bar{x} + H^T R^{-1}(m - H\bar{x})$$

whence

(2.7) $$\hat{x} = \bar{x} + P H^T R^{-1}(m - H\bar{x})$$

where

(2.8) $$P = [M^{-1} + H^T R^{-1} H]^{-1}.$$

The matrix P has an important property. To show this we determine the covariance matrix of the error $\hat{x} - x$ after measurement. We have from Eq. (2.7)

$$\hat{x} - x = \bar{x} - x + \hat{x} - \bar{x} = \bar{x} - x + P H^T R^{-1}[v - H(\bar{x} - x)]$$

or

$$\hat{x} - x = (I - PH^TR^{-1}H)(\bar{x} - x) + PH^TR^{-1}v.$$

Multiplying by the transpose and remembering that $\bar{x} - x$ and v are uncorrelated we find, with $R^T = R$,

$$E\{(\hat{x} - x)(\hat{x} - x)^T\} = (I - PH^TR^{-1}H)M(I - PH^TR^{-1}H)^T + PH^TR^{-1}HP^T.$$

But, from Eq. (2.8),

$$P^{-1} = M^{-1} + H^TR^{-1}H$$

or, after premultiplication by P,

$$PM^{-1} = I - PH^TR^{-1}H.$$

Hence, since $M^T = M$,

$$E\{(\hat{x} - x)(\hat{x} - x)^T\} = PM^{-1}P^T + (I - PM^{-1})P^T = P^T = P.$$

The $(n \times n)$-matrix P is therefore the covariance matrix of the error $\hat{x} - x$, that is, the error covariance matrix <u>after</u> measurement.

Assuming the estimate $\bar{x}$ to be the unconditioned expectation of

$$E\{x\} = \bar{x}$$

one might surmise that the estimate $\hat{x}$ represents the conditional expectation

(2.9) $$E\{x|m\} = \hat{x}.$$

This is indeed true if x and v are Gaussian. For a proof we determine the conditional density $p(x|m)$ using

$$p(x|m) = \frac{p(x,m)}{p(m)}.$$

But from Eq. (2.3)

$$p(x,m) = p(x,v) = p(x)p(v) = p(x)p(m - Hx)$$

since x and v are independent. Hence,

(2.10) $$p\{x|m\} = \frac{p(x)p(m - Hx)}{p(m)}.$$

Now

(2.11) $$\begin{cases} E\{x\} = \bar{x}, & E\{v\} = 0, & \text{Var}\{x\} = M \\ \text{Var}\{v\} = R, & E\{m\} = H\bar{x}, & \text{Var}\{m\} = HMH^T + R. \end{cases}$$

Substituting this into Eq. (2.10) and using Gauss' law we get, after some manipulation and rearranging,

(2.12) $$p(x|m) = \frac{1}{(2\pi)^{n/2}} \sqrt{\frac{|HMH^T + R|}{|M||R|}} \exp\left\{-\frac{1}{2}(x - \hat{x})^T \hat{P}^{-1}(x - \hat{x})\right\}$$

where $\hat{x}$ and P are defined by Eqs. (2.7) and (2.8). This proves Eq. (2.9).

We note that in the Gaussian case we get, through Eq. (2.12), complete information on the estimation problem. We are thus able to use the measurement to update our knowledge of the state of the system from $p(x)$ to $p(x|m)$. The method is known as the <u>Bayesian approach</u> and may, in principle at least, be employed also in the non-Gaussian and non-linear case, where Eq. (2.3) is replaced by the general relation

$$m = h(x, v) . \qquad (2.13)$$

For details see [6].

2.2. Optimal Filtering and Prediction for Linear Multistage Systems

In the preceding section we discussed a static system, that is, a system whose state vector x did not change with time. We now turn our attention to dynamic systems. Here the state vector changes, either discontinuously in discrete steps, or continuously. We consider the discrete case first.

Let the system be described by the difference equation

$$x_{i+1} = \Phi_i x_i + w_i \qquad (i = 0, 1, \ldots, N-1). \qquad (2.14)$$

The state of the system at state i is represented by the n-vector x_i. The $(n \times n)$ transition matrix Φ_i is known. The vector w_i is the noise vector with zero mean and known crosscorrelation matrix Q_i,

$$(2.15) \qquad E\{w_i w_j^T\} = Q_i \delta_{ij} = \begin{cases} Q_i & i = j \\ 0 & i \neq j \end{cases}$$

that is, the w_i constitute a purely random sequence. Hence, the x_i in turn form a Markov sequence, cf. [1], p. 68. Furthermore, we assume the initial state of the system and the noise to be uncorrelated

$$(2.16) \qquad E\{x_0 w_i^T\} = 0 .$$

As a consequence of this relation one finds immediately

$$(2.17) \qquad E\{x_i w_i^T\} = 0$$

for any i. This follows from Eq. (2.14) by postmultiplying both sides by w_j^T and taking expectations. Then

$$E\{x_{i+1} w_j^T\} = \Phi_i E\{x_i w_j^T\} + Q_i \delta_{ij} .$$

Starting this recursion formula with Eq. (2.16) one gets Eq. (2.17).

Now, while the system is at state i, measurements m_i are made which are related to the state as in Eq. (2.3):

$$m_i = H_i x_i + v_i \qquad (2.18)$$

where v_i, too, is purely random with zero expectation and given crosscorrelation

$$E\{v_i v_j^T\} = R_i \delta_{ij} . \qquad (2.19)$$

We will also assume that the two driving forces w_i and v_i are uncorrelated

$$E\{w_i v_j^T\} = 0 \quad \text{and} \quad E\{x_o v_j^T\} = 0 . \qquad (2.20)$$

State of the system and measurement noise are then also uncorrelated

$$E\{x_i v_j^T\} = 0 \qquad (2.21)$$

For an optimal estimate $\hat{x}_i$ of the system vector x_i at state i we now use the measurement vectors $m_o, m_1, \ldots, m_k$ where $k \le N$. If $i < k$ we speak of <u>smoothing</u>, for $i = k$ we have <u>filtering</u>, and for $i > k$ we have <u>prediction</u>. The case of smoothing will not be treated here. The reader is referred to [6], Chapter 13, for instance.

We use Eqs. (2.7) and (2.8)

$$\hat{x}_i = \bar{x}_i + K_i(m_i - H_i \bar{x}_i) \qquad (2.22a)$$

(2.22b) $\quad K_i = P_i H_i^T R_i^{-1}$, $\qquad P_i = [M_i^{-1} + H_i^T R_i^{-1} H_i]^{-1}$.

To these equations we adjoin the difference equation (2.14). By taking estimates $\bar{x}_i$ on both sides and replacing them on the right-hand side by the improved estimate $\hat{x}_i$ we get

(2.23) $\qquad \bar{x}_{i+1} = \Phi_i \hat{x}_i$.

We note that, as a consequence of this relation, $\bar{x}_i$ no longer represents the expectation of x_i.

We also need an expression for the covariance M_i at state i. Multiplying the equation

$$x_{i+1} - \bar{x}_{i+1} = \Phi_i (x_i - \hat{x}_i) + w_i$$

by its transposed, taking expectations, using Eqs. (2.15) and (2.17) and remembering that

$$E\{(x_i - \bar{x}_i)(x_i - \bar{x}_i)^T\} = M_i , \quad E\{(x_i - \hat{x}_i)(x_i - \hat{x}_i)^T\} = P_i$$

we obtain

(2.24) $\qquad M_{i+1} = \Phi_i P_i \Phi_i^T + Q_i$

Eqs. (2.22), (2.23) and (2.24) represent the <u>Kalman filter</u>. For a rigorous derivation see [3]. Unknown in these equations at stage i are $\bar{x}_i$, $\hat{x}_i$, M_i, P_i.

Optimal Filtering and Prediction for Linear Multistage Systems

For <u>prediction</u> to a state $j > k$ where no measurements beyond k are available we have only Eq. (2.23) at our disposal. In other words, we have to be satisfied with the estimate

$$\hat{x}_{i+1} = \bar{x}_{i+1} = \Phi_i \hat{x}_i \quad \text{for} \quad i = k, k+1, \ldots, j-1. \quad (2.25)$$

We add a few remarks.

(a) <u>Stationary System</u>. If Φ_i and H_i as well as Q_i and R_i are all constant matrices, i.e., independent of i, the filtering process may converge asymptotically to a steady state such that M_i and P_i become constant matrices as $i \to \infty$. The limiting values, if they exist, follow from Eqs. (2.22) and (2.24) as

$$\left. \begin{array}{c} P = [M^{-1} + H^T R^{-1} H]^{-1} \\ \\ M = \Phi P \Phi^T + Q \end{array} \right\}. \quad (2.26)$$

(b) <u>Gaussian noise</u>. If the purely random noise sequence w_i is Gaussian the state vector x_i, being the output of a linear system, will also be Gaussian.

(c) <u>Shaping filter</u>. Suppose that Eqs. (2.14) and (2.18) are still valid, but that v_i is now not a purely random sequence but is generated by the multistage shaping filter

$$v_{i+1} = \Psi_i v_i + \omega_i$$

where ω_i is a Gaussian random sequence with zero mean and covariance

$$E\{\omega_i \omega_j^T\} = S_i \delta_{ij}.$$

The problem may be attacked by combining x_i and v_i into a single larger state vector (augmented state)

$$y_i = \begin{pmatrix} x_i \\ v_i \end{pmatrix}.$$

In a similar manner, the corresponding matrices are combined. The method may, however, lead to computational difficulties. An improved method is described in [6], p. 402.

(d) <u>Suboptimal Filter</u>. The use of the Kalman filter requires the knowledge of the covariance matrices Q_i and R_i of the noise, and of P_0 of the initial error. Usually, only rough estimates of these matrices are available. Therefore, the filter must be designed so as to tolerate variations in Q_i and R_i. This leads to the concept of sub-optimal filters. For details see [5], p. 165.

(e) <u>Divergence</u>. Theoretically, the Kalman filter, if it is stable, should produce an increasingly accurate estimate as time proceeds and additional measurements become available. Estimation errors observed in practical situations, however, frequently tend to be much higher than predicted by theory. They may even increase monotonically. It seems that this is due to the unavoidable discrepancy between reality and model. Noise is never completely known, and system properties may vary slowly with time during each stage instead of being constant as is assumed in the theory. Various suggestions have been made recently to remedy this situation, see, for instance, [8].

Example

A very special case of Eqs. (2.14) and (2.18) is given by

$$\left.\begin{aligned} x_{i+1} &= x_i \\ m_i &= x_i + v_i \end{aligned}\right\} \quad (a)$$

This is equivalent to determining a constant vector x, or, as we shall assume here, a constant number x, by repeated measurements.

We have now

$$E\{v_i v_j\} = r\delta_{ij} , \qquad E\{(\hat{x}_i - x_i)^2\} = p_i$$

where r and p_i are numbers. Eqs. (2.23) and (2.24) give

$$\bar{x}_{i+1} = \hat{x}_i , \qquad M_{i+1} = P_i ..$$

Hence, Eqs. (2.22), with i replaced by $i+1$, read

$$\hat{x}_{i+1} = \hat{x}_i + \frac{p_{i+1}}{r}(m_{i+1} - \hat{x}_i) \qquad (b')$$

(b'')
$$\frac{1}{p_{i+1}} = \frac{1}{p_i} + \frac{1}{r} .$$

The solution of Eq. (b'') is

(c)
$$\frac{p_i}{r} = \frac{p_0}{r + i p_0} .$$

Substituting this into Eq. (b') and rearranging we find

$$[r + (i+1)p_0]\hat{x}_{i+1} = (r + i p_0)\hat{x}_i + p_0 m_{i+1} .$$

This recursive relation yields

(d)
$$(r + i p_0)\hat{x}_i = r\hat{x}_0 + p_0(m_1 + m_2 + \ldots + m_i) .$$

In the absence of measurement noise, $r = 0$, the best estimate is of course the mean

$$\hat{x}_i = \frac{1}{i}(m_1 + m_2 + \ldots + m_i) .$$

On the other hand, as the number of measurements increases, $i \to \infty$, Eqs. (c) and (d) become

$$\lim p_i = 0 \quad , \quad \lim \hat{x}_i = \lim \frac{1}{i}(m_1 + m_2 + \ldots + m_i) .$$

The error variance tends to zero and the best estimate tends to the average of all measurements, as was to be expected.

2.3. Optimal Filtering and Prediction for Linear Continuous Systems

Let the system now be described by the linear differential equation.

$$\dot{x} = F(t)x + w(t). \tag{2.27}$$

The n-vector $x(t)$ represents the state of the system. $F(t)$ is a known matrix, $w(t)$ is a white noise vector. Hence, $x(t)$ represents a Markov process.

Measurements $m(t)$, corrupted by white noise $v(t)$, are continuously made:

$$m(t) = H(t)x(t) + v(t). \tag{2.28}$$

Again, the $(p \times n)$-matrix $H(t)$ is known. w and v are uncorrelated with

$$E\{w(t)\} = 0 \quad , \quad E\{v(t)\} = 0 \tag{2.29}$$

and

$$\left. \begin{array}{l} E\{w(t)w(\tau)^T\} = Q(t)\delta(t-\tau) \\ \\ E\{v(t)v(\tau)^T\} = R(t)\delta(t-\tau) \\ \\ E\{w(t)v^T(\tau)\} = 0 \end{array} \right\}. \tag{2.30}$$

The measurements $m(\tau)$ for $t_0 \le \tau \le t_1$ will now be employed to obtain a conditional estimate $\hat{x}[t|z(t_0)...z(t_1)]$ or, briefly, $\hat{x}(t)$, for $x(t)$. If $t < t_1$ we call the procedure <u>smoothing</u>, for $t = t_1$ we have <u>filtering</u>, and for $t > t_1$ we speak of <u>prediction</u>. Again, smoothing will not be treated here. The reader is referred to [6], Chapter 13.

The corresponding filter equations may be obtained from those of the discrete case of the preceding section by a limiting procedure [6]. Due to the fact that Eq. (2.27) contains white noise and is, therefore, to be interpreted in the sense of the Ito calculus, the derivation is purely formal. The rigorous derivation may be found in [4] and [5].

We identify x_i with $x(t_i)$ and write $t_{i+1} - t_i = \theta$. In the limit, as $\theta \to 0$, the difference equations of the preceding section go over into differential equations.

<u>Eq. (2.14)</u>: In order to have Eq. (2.14) go over into Eq. (2.27) we write

$$\frac{x_{i+1} - x_i}{\theta} = \frac{\Phi_i - I}{\theta} x_i + \frac{1}{\theta} w .$$

Hence, as $\theta \to 0$, we must have

(2.31) $\qquad \lim \frac{\Phi_i - I}{\theta} = F(t) \quad , \quad \lim \frac{1}{\theta} w_i = w(t) .$

<u>Eqs. (2.22)</u>: In order to transfer Eq. (2.19) into Eq. $(2.30)_2$,

Optimal Filtering and Prediction for Linear Continuous Systems

with $\lim \frac{\delta_{ij}}{\theta} = \delta(t-\tau)$, we let

$$\lim \theta R_i = R(t) . \qquad (2.32)$$

Then

$$\lim (\hat{x}_i - \bar{x}_i) = \lim P_i H_i^T (\theta R_i)^{-1}(m_i - H\bar{x}_i)\theta = 0$$

$$\lim (P_i^{-1} - M_i^{-1}) = \lim H_i^T(\theta R_i)^{-1} H_i \theta = 0$$

i.e.,

$$\left. \begin{array}{l} \lim \hat{x}_i = \lim \bar{x}_i = \hat{x}(t) \\ \\ \lim P_i = \lim M_i = P(t) \end{array} \right\} . \qquad (2.33)$$

<u>Eq. (2.23) and (2.24)</u>: Employing Eqs. (2.33) we have from Eq. (2.23)

$$\frac{\hat{x}_{i+1} - \hat{x}_i}{\theta} = \frac{\Phi_i - I}{\theta} \hat{x}_i + \frac{\hat{x}_{i+1} - \bar{x}_{i+1}}{\theta} = \qquad (a)$$

$$= \frac{\Phi_i - I}{\theta} \hat{x}_i + P_{i+1} H_{i+1}^T (\theta R_{i+1})^{-1}(m_{i+1} - H_{i+1}\bar{x}_{i+1})$$

and from Eq. (2.24)

$$\frac{M_{i+1} - P_i}{\theta} = \frac{1}{\theta}(\Phi_i P_i \Phi_i^T - P_i) + \frac{Q_i}{\theta} = \qquad (b)$$

(b) $\qquad = \dfrac{\Phi_i - I}{\theta} P_i + P_i \dfrac{\Phi_i^T - I}{\theta} + \dfrac{\Phi_i - I}{\theta} P_i \dfrac{\Phi_i^T - I}{\theta} \theta + \dfrac{Q_i}{\theta}$.

From Eq. (2.22b), after division by θ and expansion in a power series,

$$\dfrac{P_i}{\theta} = \left[\left(\dfrac{M_i}{\theta}\right)^{-1} + \theta^2 H_i^T (\theta R_i)^{-1} H_i \right]^{-1} =$$

$$= \dfrac{M_i}{\theta} - \theta^2 \dfrac{M_i}{\theta} H_i^T (\theta R_i)^{-1} H_i \dfrac{M_i}{\theta} + O(\theta) .$$

Replacement of i by $i+1$ and substitution into Eq. (b) renders

(c) $\qquad \dfrac{P_{i+1} - P_i}{\theta} = \dfrac{\Phi_i^T - I}{\theta} P_i + P_i \dfrac{\Phi_i^T - I}{\theta} +$

$\qquad\qquad\qquad + \dfrac{Q_i}{\theta} - M_{i+1} H_{i+1}^T (\theta R_{i+1})^{-1} H_{i+1} M_{i+1} + O(\theta)$.

Upon comparing Eqs. (2.15) and (2.30), together with Eq. (2.31)$_2$, we may write

$$\lim E\left\{ \dfrac{w_i}{\theta} \dfrac{w_j^T}{\theta} \right\} = \lim \dfrac{Q_i}{\theta} \lim \dfrac{\delta_{ij}}{\theta} = Q(t) \delta(t - \tau) .$$

Hence, we have

(2.34) $\qquad\qquad\qquad \lim \dfrac{Q_i}{\theta} = Q(t)$.

Optimal Filtering and Prediction for Linear Continuous Systems

Letting now $\theta \to 0$ in Eqs. (a) and (c), and taking Eqs. (2.31), (2.32), (2.33) and (2.34) into consideration, we obtain

$$\dot{\hat{x}} = F\hat{x} + PH^TR^{-1}(m - H\hat{x})$$

$$\dot{P} = FP + PF^T + Q - PH^TR^{-1}HP \qquad (2.35)$$

with initial conditions $\hat{x}(t_o) = 0$, $P(t_o) = P_o$.

Eqs. (2.35) for the estimate $x(t)$ and the matrix $P(t)$ represent the equations of the <u>Kalman-Bucy</u> filter and are the continuous counterpart of Eqs. (2.22). We remember that $P(t) = E\{(x - \hat{x})(x - \hat{x})^T\}$ is the conditional error covariance matrix of the process.

As may be seen from a comparison of Eqs. (2.27) and (2.35) the optimal filter is obtained by taking a copy of the system and feeding the difference between the actual measurement m and the expected measurement $H\hat{x}$ into it after multiplication by a matrix K, called the <u>gain</u>

$$K = PH^TR^{-1}. \qquad (2.36)$$

The second of Eqs. (2.35) represents a system of nonlinear differential equations of the first order for the conditional error covariance matrix $P(t)$. It is known as the <u>variance equation</u> and is of the <u>Riccati</u> type.

The flow chart of the filter is shown on the next page.

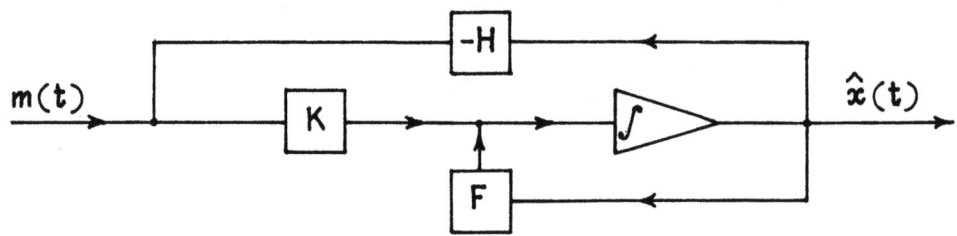

The variance equation is solved first and the resulting value of $K(t)$ is stored in the computer.

For prediction to a state $x(t)$ for $t > t_1$, i.e., beyond the time where measurements are available. Eqs. (2.35) are replaced by

$$(2.37) \quad \begin{cases} \dot{\hat{x}} = F\hat{x} \\ \dot{P} = FP + PF^T + Q \end{cases} \quad \text{for} \quad t > t_1$$

with $\hat{x}(t_1)$ and $P(t_1)$ known from Eqs. (2.35).

A few remarks follow.

(a) <u>Stationary systems</u>. If F, H, Q and R are all constant, the estimation process may reach a steady state such that P becomes constant. Its value may be determined by setting $\dot{P} = 0$ in Eq. $(2.35)_2$

$$(2.38) \quad PH^TR^{-1}HP - FP - PF^T = Q.$$

The direct solution of the $n(n+1)/2$ simultaneous quadratic equations (2.38) may be impracticable for $n > 2$; instead, one

Optimal Filtering and Prediction for Linear Continuous Systems

integrates Eq. (2.35)$_2$ with $P(0) = P_0$, by the Runge-Kutta method until $\dot{P} \approx 0$. See also [5], p. 108.

The steady-state filter is the case that we have discussed in the Wiener-Hopf theory, Chapter 1. Rewriting Eq. (2.35)$_1$ in the form

$$\dot{\hat{x}} = (F - KH)\hat{x} + Km$$

we see immediately that the impulse response matrix of the Wiener filter is given by

$$h(t) = \exp[(F - KH)t]$$

The relation between estimate $\hat{x}$ and measurement m is then

$$\hat{x}(t) = K \int_0^\infty h(\tau) m(t-\tau) d\tau.$$

The Wiener derivation, however, does not require the white-noise assumption for w and v.

(b) Colored noise. The Kalman-Bucy filter, Eqs. (2.35), is based on the assumption that the measurement noise, $v(t)$, is white and nonzero. If $v(t)$ were identically zero, i.e., if the measurement were perfect, $R(t)$ would be a null matrix and the variance equation would be singular. Also, if $v(t)$ were nonwhite, the augmented state approach, using shaping filters, could not be used, since it too would lead to a singular variance equation. To see this, consider the system

(a) $$\begin{cases} \dot{x} = F(t)x + w(t) \\ m(t) = H(t)x + v(t) \\ \dot{v} = A(t)v + z(t) \end{cases}$$

with

(b) $$\begin{cases} E\{w(t)w^T(\tau)\} = Q(t)\delta(t-\tau) \\ E\{w(t)z^T(\tau)\} = 0 \\ E\{z(t)z^T(\tau)\} = S(t)\delta(t-\tau) \end{cases}$$

and write

$$\dot{y} = \Phi(t)y + \eta$$

where

$$y = \begin{pmatrix} x \\ \hline v \end{pmatrix}, \quad \Phi = \begin{pmatrix} F & 0 \\ \hline 0 & A \end{pmatrix}, \quad \eta = \begin{pmatrix} w \\ \hline z \end{pmatrix}.$$

It follows that the measurements $m(t)$ in terms of the new state vector $y(t)$ are perfect:

$$m = (H \mid I)y.$$

Hence, $R(t) \equiv 0$.

A different procedure consists in eliminating v

Optimal Filtering and Prediction for Linear Continuous Systems

from the last two of the equations (a) by introducing a new measurement vector $\mu(t)$, defined as

$$\mu(t) = \dot{m} - A(t)m = H x + H\dot{x} + \dot{v} - Am =$$

$$= \dot{H}x + HFx + Hw + A(m - Hx) + z - Am.$$

We obtain the two equations

$$\left. \begin{array}{l} \dot{x} = Fx + w \\ \mu = H^* x + \omega \end{array} \right\} \quad (c)$$

where

$$\left. \begin{array}{l} H^* = \dot{H} + HF - AH \\ \omega(t) = Hw + z \end{array} \right\} \quad (d)$$

Formally, Eqs. (c) have the appearance of Eqs. (2.27) and (2.28). However, process noise w and measurement noise ω are now correlated :

$$E\{w(t)\omega^T(\tau)\} = E\{w(t)w^T(\tau)H^T + w(t)z^T(\tau)\} = QH^T\delta(t-\tau)$$

Bryson and Johansen [9] have solved the filtering problem for the general case. The restricted problem given above is treated in [6], p. 405.

Example

Consider the damped oscillator with one degree of

freedom and white noise excitation [6],

(a) $$\ddot{x} + 2\zeta\omega_o\dot{x} + \omega_o^2 x = u(t)$$

with

(b) $$E\{u(t)\,u(\tau)\} = q\delta(t-\tau).$$

Introducing the state vector $x_1 = x$, $x_2 = \dot{x}$ we rewrite Eq. (a) as

(c) $$\begin{pmatrix}\dot{x}_1 \\ \dot{x}_2\end{pmatrix} = \begin{pmatrix} 0 & 1 \\ -\omega_o^2 & -2\zeta\omega_o \end{pmatrix}\begin{pmatrix} x_1 \\ x_2 \end{pmatrix} + \begin{pmatrix} 0 \\ 1 \end{pmatrix} u(t).$$

A continuous measurement of the velocity x_2 is made which is corrupted by another white noise process $v(t)$, uncorrelated with $u(t)$,

(d) $$m(t) = (0\quad 1)\begin{pmatrix} x_1 \\ x_2 \end{pmatrix} + v(t)$$

with

(e) $$E\{v(t)\,v(\tau)\} = r\delta(t-\tau).$$

Eqs. (2.35) of the Kalman-Bucy filter reduce here to

$$\begin{pmatrix}\dot{\hat{x}}_1 \\ \dot{\hat{x}}_2\end{pmatrix} = \begin{pmatrix} 0 & 1 \\ -\omega_o^2 & -2\zeta\omega_o \end{pmatrix}\begin{pmatrix} \hat{x}_1 \\ \hat{x}_2 \end{pmatrix} + \begin{pmatrix} P_{11} & P_{12} \\ P_{12} & P_{22} \end{pmatrix}\begin{pmatrix} 0 \\ 1 \end{pmatrix}\frac{1}{r}(m - \hat{x}_2)$$

$$\begin{pmatrix} \dot{P}_{11} & \dot{P}_{12} \\ \dot{P}_{12} & \dot{P}_{22} \end{pmatrix} = \begin{pmatrix} 0 & 1 \\ -\omega_0^2 & -2\zeta\omega_0 \end{pmatrix} \begin{pmatrix} P_{11} & P_{12} \\ P_{12} & P_{22} \end{pmatrix} + \begin{pmatrix} P_{11} & P_{12} \\ P_{12} & P_{22} \end{pmatrix} \begin{pmatrix} 0 & -\omega_0^2 \\ 1 & -2\zeta\omega_0 \end{pmatrix} +$$

$$+ \begin{pmatrix} 0 \\ 1 \end{pmatrix} q (0 \ 1) - \begin{pmatrix} P_{11} & P_{12} \\ P_{12} & P_{22} \end{pmatrix} \begin{pmatrix} 0 \\ 1 \end{pmatrix} \frac{1}{r} (0 \ 1) \begin{pmatrix} P_{11} & P_{12} \\ P_{12} & P_{22} \end{pmatrix}$$

or

$$\begin{pmatrix} \dot{\hat{x}}_1 \\ \dot{\hat{x}}_2 \end{pmatrix} = \begin{pmatrix} 0 & 1 \\ -\omega_0^2 & -2\zeta\omega_0 \end{pmatrix} \begin{pmatrix} \hat{x}_1 \\ \hat{x}_2 \end{pmatrix} + \begin{pmatrix} P_{12} \\ P_{22} \end{pmatrix} \frac{m\,\hat{x}_2}{r} \quad (f)$$

$$\begin{pmatrix} \dot{P}_{11} & \dot{P}_{12} \\ \dot{P}_{12} & \dot{P}_{22} \end{pmatrix} = \begin{pmatrix} 2P_{12} & P_{22} - \omega_0^2 P_{11} - 2\zeta\omega_0 P_{12} \\ P_{22} - \omega_0^2 P_{11} - 2\zeta\omega_0 P_{12} & -2\omega_0^2 P_{12} - 4\zeta\omega_0 P_{22} \end{pmatrix}$$

$$- \frac{1}{r} \begin{pmatrix} P_{12}^2 & P_{12} P_{22} \\ P_{12} P_{22} & P_{22}^2 \end{pmatrix} + \begin{pmatrix} 0 & 0 \\ 0 & 1 \end{pmatrix} q \ .$$

(g)

The initial values of $\hat{x}_1$, $\hat{x}_2$ and of $P_{11}(0)$, $P_{12}(0)$, $P_{22}(0)$ are assumed to be given (equal to zero, for instance).

The three coupled Riccati equations (g) can only be solved numerically. The resulting $P_{12}(t)$ and $P_{22}(t)$ are then used to solve Eq. (f) for the best estimate $\hat{x}_1(t)$ and $\hat{x}_2(t)$.

Putting $\dot{P}_{11} = \dot{P}_{12} = \dot{P}_{22} = 0$ in Eqs. (g) one finds for the steady-state covariances

(h) $$\begin{cases} P_{11} = 2\zeta \dfrac{r}{\omega_o} \left[\sqrt{1 + \dfrac{1}{4\zeta^2 \omega_o^2} \dfrac{q}{r}} - 1 \right] \\ P_{22} = \omega_o^2 P_{11}, \quad P_{12} = 0. \end{cases}$$

Chapter 3
OPTIMAL FILTERING FOR NONLINEAR SYSTEMS

Both the Wiener and the Kalman filter discussed in the preceding Chapters have been designed for linear systems. Real systems, however, are mostly nonlinear. It is, therefore, essential to have suitable linearizing procedures available, so that the filters may still be applied.

Two groups of methods may be distinguished. In the first, system and measurement equations are linearized and the filter equations are then applied. In the second, the filter equations themselves are partly linearized.
We discuss the methods briefly in the following.

(a) Linearization of System and Measurements About a Nominal Path

Let system and measurements be governed by a set of nonlinear difference equations:

$$\left.\begin{array}{l} x_{i+1} = \varphi_i(x_i, W_i) \\ \\ m_i = h_i(x_i, V_i) \end{array}\right\} \quad (3.1)$$

where x_i and m_i have the same meaning as in Section 2 of Chapter 2, and W_i and V_i are purely random sequences with zero expectations. Select for the nominal path a sequence ζ_i of deterministic state vectors, satisfying any initial and final conditions that may be imposed on x_i. Now expand in a Taylor series about the nominal path and about $W_i = V_i = 0$, retaining only terms up to the first order (*), that is,

$$(3.2) \quad \begin{cases} x_{i+1} = \varphi_i(\xi_i, 0) + \Phi_i[x_i - \xi_i] + w_i \\ m_i = h_i(\xi_i, 0) + H_i[x_i - \xi_i] + v_i \end{cases}$$

where

$$(3.3) \quad \Phi_i = \frac{\partial \varphi_i}{\partial x_i}\bigg|_{x_i = \xi_i, W_i = 0} , \quad H_i = \frac{\partial h_i}{\partial x_i}\bigg|_{x_i = \xi_i, V_i = 0}$$

$$u_i = \frac{\partial \varphi_i}{\partial W_i}\bigg|_{x_i = \xi_i, W_i = 0} , \quad v_i = \frac{\partial h_i}{\partial V_i}\bigg|_{x_i = \xi_i, V_i = 0}.$$

Putting

$$(3.4) \quad \alpha_i = x_i - \xi_i, \quad \beta_{i+1} = x_{i+1} - \varphi_i(\xi_i, 0), \quad \gamma_i = m_i - h_i(\xi_i, 0).$$

Eqs. (3.2) take on the form of Eqs. (2.14) and (2.18):

$$(3.5) \quad \beta_{i+1} = \Phi_i \alpha_i + w_i$$

$$\gamma_i = H_i \alpha_i + v_i.$$

(*) cf. [10] p. 180, [7] p. 512 and [5] p. 143.

We see, then, that the corresponding Kalman filter is given by

$$\hat{\beta}_i = \bar{\beta}_i + K_i[\hat{y}_i - H_i\bar{\beta}_i] \qquad (3.6)$$

where

$$\bar{\beta}_{i+1} = \Phi_i \hat{\alpha}_i \qquad (3.7)$$

and K_i is the gain and where it has been assumed that the same covariances hold for w_i and v_i as in Section 2 of Chapter 2.

The corresponding approximately optimal estimate of the state vector is

$$\hat{x}_{i+1} = \hat{\beta}_{i+1} + \varphi_i(\xi_i, 0) . \qquad (3.8)$$

Furthermore,

$$\hat{\alpha}_i = \hat{x}_i - \xi_i . \qquad (3.9)$$

If the nominal sequence ξ_i can be chosen to satisfy the nonlinear system equations with noise absent, then $\alpha_i = \beta_i$. A procedure, which might be useful in achieving this is quasilinearization.

Quasilinearization. The basic idea of quasilinearization may easily be explained . Suppose we have a set of nonlinear equations of first order

$$\dot{x} = f(x, t) \qquad (3.10)$$

where x is a n-vector. The solution of Eq. (3.10) is to satisfy certain boundary conditions, i.e., initial conditions at $t = t_0$ for one part of the components, and final conditions at $t = t_f$ for the remaining components. Expanding now $f(x,t)$ about the kth approximation $x^{(k)}(t)$ of $x(t)$, we replace Eq. (3.10) by the following differential equation

$$(3.11) \quad \dot{x}^{(k+1)} = \frac{\partial f^{(k)}}{\partial x} [x^{(k+1)} - x^{(k)}] + f(x^{(k)}, t)$$

where

$$(3.12) \quad \frac{\partial f^{(k)}}{\partial x} = \frac{\partial f(x,t)}{\partial x} \bigg|_{x = x^{(k)}}.$$

The initial approximation $x^{(0)}(t)$ may be any function that satisfies the boundary conditions.

If $|x^{(k+1)} - x^{(k)}| \leq \varepsilon$, and $\varepsilon > 0$ is sufficiently small, the differential equation (3.11) is a sufficiently close approximation to Eq. (3.10). Sufficient conditions for the convergence of the procedure have been given by Kalaba and may be found in [11].

The modification of the method for a difference equation is obvious and may be applied to the problem of the previous section.

A different combination of quasilinearization with Kalman filtering is presented by Greensite, [7], p. 533, for a very special case.

(b) Partial Linearization of the Filter Equations

The method is due to <u>Bryson and Ho</u>. We restrict attention here to the case of a nonlinear continuous system. The equations for the multi-stage system, as well as many additional details may be found in [6], p. 373.

The nonlinear system is assumed to be disturbed by additive white noise

$$\dot{x} = f(x, t) + w(t). \qquad (3.13)$$

The same holds true for the continuous measurements

$$m(t) = h(x, t) + v(t). \qquad (3.14)$$

It is assumed that relations (2.29) and (2.30) for expectations and covariances, together with

$$\left.\begin{array}{l} E\{[x(t_0) - \bar{x}_0][x(t_0) - \bar{x}_0]^T\} = P_0 \\[6pt] E\{[x(t_0) - \bar{x}_0]w(t)^T\} = 0 \\[6pt] E\{[x(t_0) - \bar{x}_0]v(t)^T\} = 0 \end{array}\right\} . \qquad (3.15)$$

hold.

Eqs. (2.34) are rewritten in the form

$$\dot{\hat{x}} = f(\hat{x}, t) + P\left(\frac{\partial h}{\partial x}\right)^T R^{-1}[m(t) - h(\hat{x}, t)], \quad \hat{x}(t_0) = \bar{x}_0 \qquad (3.16a)$$

$$(3.16\mathrm{b}) \quad \dot{P} = \frac{\partial f}{\partial x}P + P\left(\frac{\partial f}{\partial x}\right)^T + Q - P\left(\frac{\partial h}{\partial x}\right)^T R^{-1}\frac{\partial h}{\partial x}P, \quad P(t_o) = P_o \ .$$

The derivatives are evaluated along a preselected nominal path, or, for greater accuracy, with $\xi = \hat{x}$, the latter case corresponding to a continuous relinarization. $P(t)$ is then coupled to the current estimate $\hat{x}(t)$ and cannot be precalculated as in the former case.

Example

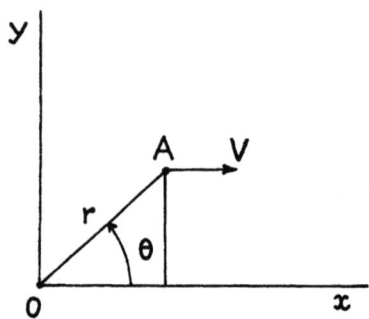

An airplane A is flying horizontally at constant speed V over a flat earth. It is being tracked from a fixed point O. Measurements of distance r and elevation θ are continuously taken to determine distance x and altitude y. The path of the airplane is disturbed by air turbulence and the measurements are contaminated by noise. Both disturbances are white with zero expectation.

The equations of motion are

(a) $$\begin{cases} \dot{x} = V + w_1 \\ \dot{y} = w_2 \end{cases}$$

Optimal Filtering for Nonlinear Systems

and for the measurements we have

$$\left.\begin{array}{l} m_1 = \rho = r + v_1 = \sqrt{x^2 + y^2} + v_1 \\[2mm] m_2 = \vartheta = \theta + v_2 = \arctan \dfrac{y}{x} + v_2 \end{array}\right\} . \quad (b)$$

We use method (b) and evaluate equations (3.16) along the nominal path $\xi = Vt$, $\eta = \text{const}$. From Eqs. (3.13) and (3.14) we have then

$$f = \begin{pmatrix} V \\ 0 \end{pmatrix} , \quad h = \begin{pmatrix} \sqrt{x^2 + y^2} \\ \arctan \dfrac{y}{x} \end{pmatrix}$$

and hence

$$\dfrac{\partial f}{\partial \underset{\sim}{x}} = 0 , \quad \dfrac{\partial h}{\partial \underset{\sim}{x}} = \left(\dfrac{\partial h_i}{\partial x_j}\right) = \begin{pmatrix} \dfrac{x}{r} & \dfrac{y}{r} \\ -\dfrac{y}{r^2} & \dfrac{x}{r^2} \end{pmatrix} .$$

Along the reference path

$$\dfrac{\partial f}{\partial \underset{\sim}{x}} = 0 , \quad \dfrac{\partial h}{\partial \underset{\sim}{x}} = \begin{pmatrix} \dfrac{Vt}{r_n} & \dfrac{\eta}{r_n} \\ -\dfrac{\eta}{r_n^2} & \dfrac{Vt}{r_n^2} \end{pmatrix} , \quad r_n = \sqrt{(Vt)^2 + \eta^2} . \quad (c)$$

Eqs. (3.16) read now, if $E\{v_1(t)v_2(t)\} = 0$, i.e., $R_{12} = R_{21} = 0$ is assumed,

(d)
$$\begin{pmatrix}\dot{\hat{x}} \\ \dot{\hat{y}}\end{pmatrix} = \begin{pmatrix} V \\ 0 \end{pmatrix} + \begin{pmatrix} P_{11} & P_{12} \\ P_{12} & P_{22} \end{pmatrix}\begin{pmatrix} Vt & \dfrac{-\eta}{r_n} \\ \eta & \dfrac{Vt}{r_n} \end{pmatrix}$$

$$\cdot \begin{pmatrix} \dfrac{1}{R_{11}} & 0 \\ 0 & \dfrac{1}{R_{22}} \end{pmatrix} \dfrac{1}{r_n}\begin{pmatrix} \rho - \sqrt{\hat{x}^2 + \hat{y}^2} \\ \vartheta - \arctan \dfrac{\hat{y}}{\hat{x}} \end{pmatrix}.$$

and

(e)
$$\begin{pmatrix} \dot{P}_{11} & \dot{P}_{12} \\ \dot{P}_{12} & \dot{P}_{22} \end{pmatrix} = \begin{pmatrix} Q_{11} & Q_{12} \\ Q_{21} & Q_{22} \end{pmatrix} - \begin{pmatrix} P_{11} & P_{12} \\ P_{12} & P_{22} \end{pmatrix}\begin{pmatrix} Vt & \dfrac{-\eta}{r_n} \\ \eta & \dfrac{Vt}{r_n} \end{pmatrix}$$

$$\cdot \begin{pmatrix} \dfrac{1}{R_{11}} & 0 \\ 0 & \dfrac{1}{R_{22}} \end{pmatrix}\begin{pmatrix} Vt & \eta \\ -\eta & Vt \\ \hline r_n & r_n \end{pmatrix}\begin{pmatrix} P_{11} & P_{12} \\ P_{12} & P_{22} \end{pmatrix}\dfrac{1}{r_n^2}.$$

The five nonlinear equations must be solved numerically by the Runge-Kutta method. The question of what to assume for the covariance matrices R_{ij} and Q_{ij} is, of course, a different matter and will not be discussed here. Some relevant remarks may be found in [5], p. 145 and 159. See also Chapter 2, Section 2c.

Chapter 4
OPTIMAL STOCHASTIC CONTROL

So far we have applied the Kalman filter to systems which were subjected to random disturbances but were not controlled. Very briefly we turn now our attention to the case were we wish to employ measurements to control a system in some optimal manner. Only the simplest problem will be discussed here.

As an introduction we recapitulate some basic facts from deterministic optimal control theory. We restrict our attention to continuous systems.

4.1. Deterministic Optimal Control Without Constraints

Let the system be described by the vector differential equation

$$\dot{x} = f(x, u, t) \qquad (4.1)$$

where $x(t)$ is the state n-vector and $u(t)$ the control m-vector. The latter is to be chosen such as to make the performance index (cost function)

$$J = \varphi(x, t)\Big|_{t=T} + \int_0^T L(x(\tau), u(\tau), \tau)\,d\tau \qquad (4.2)$$

a minimum over the finite time interval $[0, T]$.

With the aid of the calculus of variations the

following procedure for the solution of this problem may be formulated.

Construct a scalar function

(4.3) $$\mathscr{H} = p^T f - L$$

known as the Hamiltonian, where $p(t)$ is the vector adjoint to $x(t)$, satisfying the differential equation

(4.4) $$\dot{p}^T = -\frac{\partial \mathscr{H}}{\partial x}$$

and subject to the boundary condition

(4.5) $$p^T(T) = \frac{\partial \varphi}{\partial x} .$$

Then the problem of minimizing the functional J is equivalent to maximizing the function $\mathscr{H}$:

(4.6) $$\frac{\partial \mathscr{H}}{\partial u} = 0 .$$

We specialize now to a linear system

(4.7) $$\dot{x} = F(t)x + G(t)u$$

and a quadratic performance index

(4.8) $$J = \frac{1}{2}\left[(x^T S x)\Big|_{t=T} + \int_0^T (x^T A x + u^T B u)\,d\tau \right]$$

Deterministic Optimal Control without Constraints

S, $A(t)$ and $B(t)$ are positive definite symmetric matrices.

The Hamiltonian reads here

$$\mathcal{H} = p^T(Fx + Gu) - \frac{1}{2}(x^T A x + u^T B u)$$

and we have for the adjoint vector

$$\dot{p}^T = -\frac{\partial \mathcal{H}}{\partial x} = -p^T F + x^T A \qquad (4.9)$$

with boundary condition (4.5)

$$p(T) = S x(T). \qquad (4.10)$$

The optimality condition (4.6) renders

$$\frac{\partial \mathcal{H}}{\partial u} = 0 = p^T G - u^T B$$

whence

$$u = B^{-1} G^T p . \qquad (4.11)$$

Substituting this into Eq. (4.7) we get, together with Eq. (4.9),

$$\left. \begin{array}{ll} \dot{x} = Fx + GB^{-1}G^T p , & x(0) = x_o \\ \dot{p} = Ax - F^T p & p(T) = Sx(T) \end{array} \right\} . \qquad (4.12)$$

The solution of these two linear equations is obtained by putting

(4.13) $$p(t) = S(t)x(t)$$

which already satisfies the terminal condition (4.10). Substitution of Eq. (4.13) into (4.12) yields

$$\dot{S}x + S\dot{x} = Ax - F^T S x.$$

Substitution of $\dot{x}$ from Eq. (4.12), renders

$$(\dot{S} + SF + F^T S + SGB^{-1}G^T S - A)x = 0.$$

Since $x \neq 0$, we must have

(4.14) $$\dot{S} = -SF - F^T S - SGB^{-1}G^T S + A.$$

This is a Riccati equation with $S(T)$ prescribed. We note that $S(t)$ is a symmetric matrix. Eq. (4.14) must be integrated backwards from $t = T$ to $t = 0$.

The optimal control follows now from Eqs. (4.11) and (4.13) as

(4.15) $$u(t) = C(t)x(t)$$

where

(4.16) $$C(t) = B^{-1}(t)G^T(t)S(t)$$

represents the gain of the feedback.

4.2. Optimal Control in the Presence of Noise

Instead of Eq. (4.17) the system is now governed by the differential equation

$$\dot{x} = F(t)x + G(t)u + w(t) \qquad (4.17)$$

which differs from Eq. (2.27) by the inclusion of the control term. We wish to control the system in an optimal manner utilizing measurements as in Chapter 2,

$$m(t) = H(t)x + v(t). \qquad (4.18)$$

Since $x(t)$ is now a random process we change the performance index (4.8) to

$$J = \frac{1}{2} E\left\{ (x^T S x)_{t=T} + \int_0^T (x^T A x + u^T B u) d\tau \right\}. \qquad (4.19)$$

By invoking the so-called "Certainty-Equivalence Principle" the solution to the above problem may be written down immediately. This principle states that the equations of the stochastic case are identical with those of the deterministic case treated previously provided $x(t)$ is replaced by the estimate $\hat{x}(t)$. This leads to the following set of equations

$$\left. \begin{array}{l} u(t) = C(t)\hat{x}(t) \\[2mm] \dot{\hat{x}} = F\hat{x} + Gu + K(m - H\hat{x}), \quad \hat{x}(0) \text{ given} \end{array} \right\} \qquad (4.20)$$

where

(4.21) $$\begin{cases} C = B^{-1}G^T S \\ K = PH^T R^{-1} \end{cases}$$

and $S(t)$ and $P(t)$ follow from the two Riccati equations

(4.22) $$\begin{cases} \dot{S} = -SF - F^T S + A - C^T BC, & S(T) \text{ given} \\ \dot{P} = FP + PF^T + Q - KRK^T, & P(0) \text{ given} \end{cases}$$

For a proof of these relations the reader is referred to [6], p. 415.

Since Eqs. (4.22) are independent of the measurements they can be solved (backwards and forwards, respectively) beforehand and $S(t)$ and $P(t)$ stored.

Note. A valuable contribution to the investigation of nonlinear optimal stochastic control has been given by <u>G.T. Schmidt</u> [12] in his dissertation. He considered the stochastic non-linear system

(4.23) $$\dot{x} = f(x, u, t) + w(t)$$

with measurements

(4.24) $$m = h(x, u, t) + v(t).$$

By introducing a reference noise-free system

$$\dot{\xi} = f(\xi, u^*, t) , \quad \xi(0) = E\{x(0)\}$$

and assuming that a control exists that keeps the noisy system in the neighborhood of the reference trajectory, Schmidt succeeded in finding an approximate solution to Eqs. (4.23) and (4.24) for a general performance index. The method appears to be particularly well suited for the investigation of systems containing an unknown random parameter.

References

[1] H. Parkus: Random Processes in Mechanical Sciences. International Centre for Mechanical Sciences. Udine 1969.

[2] A. A. Sweschnikow: Untersuchungsmethoden der Theorie der Zufallsfunktionen. Teubner, Leipzig 1965.

[3] R. E. Kalman: A new approach to linear filtering and prediction problems. J. Basic Engng 82(1960), 35.

[4] R. E. Kalman and R. S. Bucy: New results in linear filtering and prediction theory. J. Basic Engng 83(1961), 95.

[5] R. S. Bucy and P. D. Joseph: Filtering for Stochastic Processes with Applications to Guidance. Interscience Publishers, New York 1968.

[6] A. E. Bryson, Jr. and Yu-Chi Ho: Applied Optimal Control. Ginn and Co., Waltham 1969.

[7] A. L. Greensite: Elements of Modern Control Theory. Spartan Books, New York 1970.

[8] T. J. Tarn and J. Zaborsky: A practical, nondiverging filter. AIAA J. 8(1970), 1127, and AIAA J. 9(1971), 767.

[9] A. F. Bryson and D. E. Johansen: Linear filtering for time-varying systems using measurements containing colored noise. IEEE Trans. Aut. Control. Vol. AC-10, 1965, 4.

[10] M. Aoki: Optimization of Stochastic Systems. Academic Press, New York 1967.

[11] R. Kalaba: On nonlinear differential equations. The maximum operation and monotone convergence. J. Math. Mech. 8(1959), 519.

[12]   G.T. Schmidt: Closed-loop control of stochastic nonlinear systems. Automatica 7 (1971), 557.

Contents

	Page
Preface	3
Introduction	5
Chapter 1 : The Wiener Filter	7
1.1. The Wiener-Hopf Equation	8
1.2. Solution of the Wiener-Hopf Equation	9
Chapter 2 : The Kalman Filter	16
2.1. Best Estimate for Linear Systems	17
2.2. Optimal Filtering and Prediction for Linear Multistage Systems	21
2.3. Optimal Filtering and Prediction for Linear Continuous Systems	29
Chapter 3 : Optimal Filtering for Nonlinear Systems	41
Chapter 4 : Optimal Stochastic Control	49
4.1. Deterministic Optimal Control Without Constraints	49
4.2. Optimal Control in the Presence of Noise	53
References	57

If you have any concerns about our products,
you can contact us on
ProductSafety@springernature.com

In case Publisher is established outside the EU,
the EU authorized representative is:
**Springer Nature Customer Service Center GmbH
Europaplatz 3, 69115 Heidelberg, Germany**

Printed by Libri Plureos GmbH
in Hamburg, Germany